Mystery Forces

by Ashley Chase
illustrated by John Gordon

Published and Distributed by

These materials are based upon work partially supported by the National Science Foundation under grant number ESI-0242733. The Federal Government has certain rights in this material. Any opinions, findings, and conclusions or recommendations expressed in this material are those of the author(s) and do not necessarily reflect the views of the National Science Foundation.

© 2009 by The Regents of the University of California. All rights reserved. No part of this publication may be reproduced or transmitted in any form or by any means, electronic or mechanical, including photocopy, recording, or any information storage or retrieval system, without permission in writing from the publisher.

Developed at Lawrence Hall of Science and the Graduate School of Education at the University of California at Berkeley

Seeds of Science/Roots of Reading™ is a collaboration of a science team led by Jacqueline Barber and a literacy team led by P. David Pearson and Gina Cervetti.

Delta Education LLC
PO Box 3000
Nashua, NH 03061
1-800-258-1302
www.deltaeducation.com

Mystery Forces
594-0044
ISBN: 978-1-59821-518-2
1 2 3 4 5 6 7 8 9 10   13 12 11 10 09 08

Contents

Force Detective 4

Case of the Shrinking Apple Tree 7

Case of the Moving Spoon 9

Case of the North-pointing Needle 11

Case of the Sticky Socks 13

Case of the Runaway Car 15

Case of the Floating Train 17

Glossary 20

Force Detective

A train floats in the air. A tree shrinks instead of growing. A spoon seems to move by itself. These are some of the mysteries you will read about in this book.

In any good mystery, there are suspects. In most mysteries, suspects are the people a detective thinks might have broken the law. But the mysteries in this book are different. In *these* mysteries, the suspects are **forces**. These forces are pushes and pulls between **objects**. They can act when the objects aren't even touching each other!

The Suspects

Gravity

Gravity is a pulling force. Earth pulls on objects with the force of gravity. The objects are pulled down toward the center of the Earth.

The Moon and other objects have gravity, too. In fact, everything pulls on everything else because of gravity. But we only notice this pull when something is huge, like a **planet** or moon.

Magnetic Force

Magnets attract (pull on) some kinds of **metal**. A magnet can attract or **repel** (push against) another magnet. These forces are **magnetic forces**.

Magnets have two **poles**, north and south. The north pole of one magnet attracts the south pole of another. If you put two north poles together, they repel each other. So do two south poles.

Earth is a huge magnet, with **magnetic** poles in the north and the south. The magnetic forces of the Earth are different from the force of gravity.

Electrostatic Force

Electrostatic force is the force that acts when an object becomes **charged**. Most of the time, objects become charged when they rub against each other. Electrostatic force can make objects attract or repel each other.

You can be a force detective. Read the cases in this book. To solve each case, you will need to look for **evidence**—clues that help you **explain** what happened. Figure out which force is at work. Is it gravity, magnetic force, or electrostatic force? Then think about what the mystery force did. Turn the page to check your **explanation**.

Case of the Shrinking Apple Tree

The apple tree in the park used to be taller than the swing set. This summer, lots of apples are growing on the tree. The branches are hanging down. Now the tree is shorter than the swing set!

Trees usually grow bigger. Why is this tree shrinking?

What force is at work here? What evidence of that force can you find? What is pushing or pulling on what?

Explanation

In the case of the shrinking apple tree, the mystery force is gravity. Earth is pulling the heavy apples down with the force of gravity. The branches aren't strong enough to hold so many apples up. That's why they are hanging down. With the branches hanging down, the tree is shorter than it usually is.

Case of the Moving Spoon

A magician is sitting at a table. She stares at a metal spoon, and it slides across the table. You ask her to move a glass. She says her magic only works on metal objects.

How can the magician move the spoon without touching it? Why does her magic only work on metal objects?

What force is at work here? What evidence of that force can you find? What is pushing or pulling on what?

Explanation

In the case of the moving spoon, the mystery force is magnetic force. The magician has a strong magnet under the table. Even through the table, the magnet attracts the metal spoon. When the magician moves the magnet, the spoon moves with it. She can only move metal objects because magnets pull strongly on some kinds of metal. Magnets don't attract glass or other **materials**.

Case of the North-pointing Needle

These sailors always know whether they are going the right way. They have a needle stuck to a piece of wood. The sailors float the wood and needle in a bowl of water. The floating needle always turns until one end is pointing south and the other end is pointing north.

Why does the needle turn this way?

What force is at work here? What evidence of that force can you find? What is pushing or pulling on what?

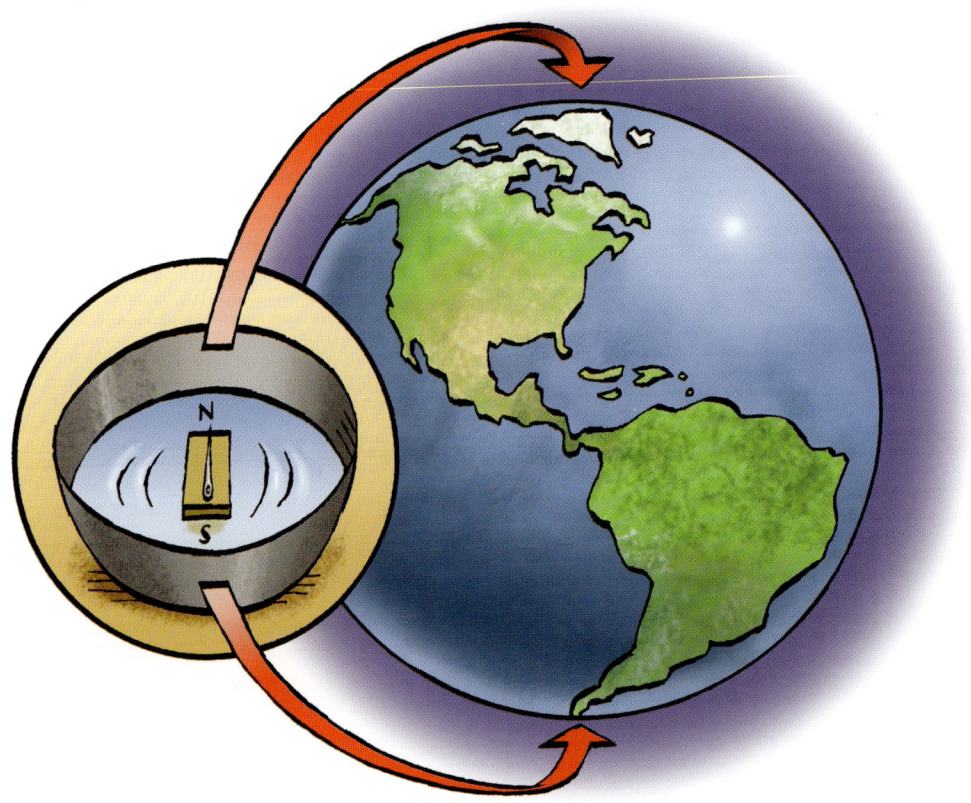

Explanation

In the case of the north-pointing needle, the mystery force is magnetic force. The sailors' needle is a **compass** needle. The needle is a magnet with a north pole at one end and a south pole at the other.

Earth is also a magnet, with magnetic poles in the north and the south. Each magnetic pole of the Earth attracts one of the poles of the sailors' compass needle. That is why one end of the needle points north and the other points south.

Case of the Sticky Socks

You pull two clean socks out of the dryer. The socks are stuck together.

Why did the socks stick together?

What force is at work here? What evidence of that force can you find? What is pushing or pulling on what?

Explanation

In the case of the sticky socks, the mystery force is electrostatic force. When the socks rolled around in the dryer, they rubbed against other clothes. One sock became charged. The socks are attracting each other.

Case of the Runaway Car

A man parks his car at the top of a hill and goes into his house. The next morning, the car is against a bush at the bottom of the hill. The car is still locked, just the way the man left it. No person has been inside the car.

How did the car move without anyone driving it?

What force is at work here? What evidence of that force can you find? What is pushing or pulling on what?

Explanation

In the case of the runaway car, the mystery force is gravity. Earth is always pulling the car and everything around it down with the force of gravity. The man parked the car at the top of a hill. He didn't set the brake that stops the car from rolling. The pull of Earth's gravity made the car roll down the hill. The car stopped when it ran into the bush.

Case of the Floating Train

This train hardly bumps as it speeds along. The ride is so smooth because the train doesn't touch the tracks. When the train moves, it floats a tiny bit above the metal tracks.

How can a train float in the air?

What force is at work here? What evidence of that force can you find? What is pushing or pulling on what?

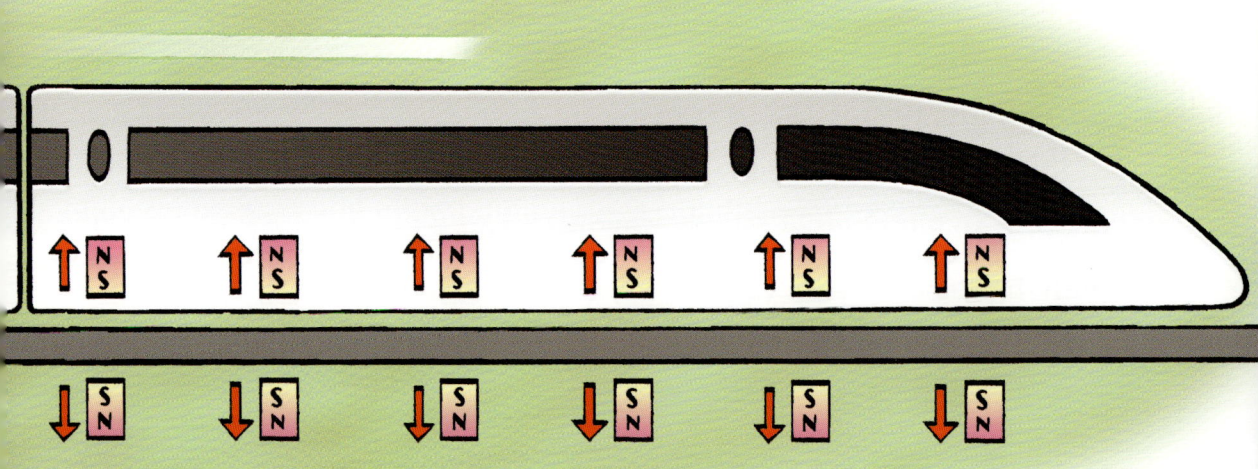

Explanation

In the case of the floating train, the mystery force is magnetic force. Strong magnets are in the tracks and the train. The magnets are placed so that matching poles are pointing toward each other. The poles of the magnets in the tracks repel the poles of the magnets in the train. The magnets repel each other strongly enough to push the heavy train into the air!

These mystery forces are around you all the time. Be a force detective in your own life. Look around for evidence of magnetic force, gravity, and electrostatic force. Figure out which force is at work when you see someone's hair sticking up or leaves falling from a tree. Before long, you'll start finding mystery forces everywhere!

Glossary

attract: to pull on something

charged: able to cause an electrostatic force

compass: a tool for figuring out direction that uses a magnet that can turn to point north and south

electrostatic force: a push or a pull between a charged object and another object

evidence: clues that help prove or explain something

explain: to give reasons for something

explanation: a description of the reasons that explain something

force: a push or a pull

gravity: a pull between objects, such as the pull between Earth and an object

magnet: something that pulls on some kinds of metal and pushes and pulls on other magnets

magnetic: having to do with magnetic force

magnetic force: the push or pull between two magnets, or the pull of a magnet on some kinds of metal

material: the stuff that things are made of

metal: a strong material that is usually very hard and shiny but can be formed into different shapes

object: a thing that can be seen or touched

planet: a huge object, such as Earth or Mars, that orbits a star

pole: the part of a magnet that pushes or pulls

repel: to push on something